ADVENTURE IN SPAC

THE FLIGHT TO FIX THE HUBBL

ELAINE SCOTT ○ MARGARET MILLER

Hyperion Books for Children
New York

Photo Cr

Pages 15,
© 1995 by
Page 52 ©
From the
All other

dits

20, 25, 27, 28, 32, 40, and 47 (bottom right)
Margaret Miller.
1993 Smithsonian Institution/Lockheed Corporation.
max film *Destiny in Space*.
photographs courtesy of NASA.

Printed in Singapore.
For information address Hyperion Books for Children,
114 Fifth Avenue, New York, New York 10011.

FIRST EDITION
1 3 5 7 9 10 8 6 4 2

Library of Congress Cataloging-in-Publication Data
Scott, Elaine.
Adventure in space: the flight to fix the Hubble/by Elaine
Scott; photographs by Margaret Miller—1st ed.
p. cm.
ISBN 0-7868-0038-0 (trade) — ISBN 0-7868-2031-4 (lib. bdg.) — ISBN 0-7868-1039-4 (pbk.)
1. Hubble Space Telescope—Maintenance and repair—Juvenile
literature. 2. Astronautics in astronomy—Juvenile literature.
[1. Hubble Space Telescope—Maintenance and repair. 2. Outer space—
Exploration.] I. Miller, Margaret, ill. II. Title.
QB500.268.S36 1994
522'.2919—dc20 94-7756 CIP AC

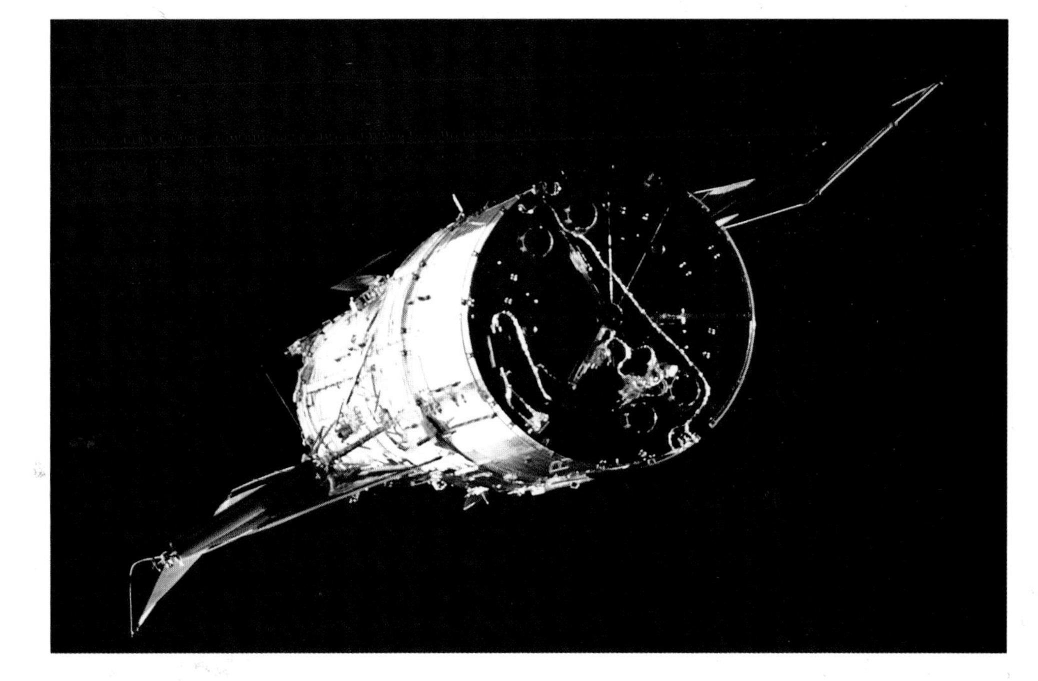

Contents

For the Adventurers—the crew of STS-61

Lieutenant Colonel Thomas Akers, United States Air Force; Commander Kenneth Bowersox, United States Navy; Colonel Richard O. Covey, United States Air Force; Jeffrey Hoffman, Ph.D.; Story Musgrave, M.D.; Captain Claude Nicollier, Swiss Air Force; and Kathryn Thornton, Ph.D.

Other desires perish in their gratification, but the desire of knowledge never: the eye is not satisfied with seeing nor the ear filled with hearing. . . . The sum of things to be known is inexhaustible, and however long we read we shall never come to the end of our storybook.

—A. E. Housman,
Selected Prose

Acknowledgments

Flying into space and writing books are similar ventures—neither could happen without the cooperation of many people. We owe a heartfelt thank-you to Brenda and Mark Thompson, who suggested the idea, introduced us to their colleagues at NASA, and provided enthusiastic support throughout our research. In addition, we are deeply grateful to the crew of STS-61—especially mission commander Dick Covey, who granted permission for this project, and mission specialists Tom Akers and Kathy Thornton, who allowed us glimpses into their private lives. The following people at the Johnson Space Center in Houston, Texas, made special arrangements for us and graciously answered myriad questions, so many thanks to Kyle Herring, Barbara Schwartz, Lucy Lytwynsky, Mary Louise Schmidt, Mike Gentry, Kara White, Chyrel Coker, Mike Fohey, and Manny Lott. Also we want to thank the Space Telescope Science Institute, IMAX Corporation, and David Reuther for his insight and advice. Lastly, we want to thank all the NASA employees at the Johnson and Kennedy space centers—too numerous to mention—who consistently answered our questions with good humor and patience.

A book must be published before it can be read. At Hyperion Books for Children we want to thank Andrea Cascardi, Ellen Friedman, Kristen Behrens, Mara Van Fleet, Connie Hahn, and Caleb Crain for their enthusiasm and commitment as we worked together to bring *Adventure in Space* to life.

INTRODUCTION

Imagine a telescope that can see back in time to the very beginnings of our universe . . . a telescope that can watch as stars are born and die, solar systems develop, and entire galaxies form. Imagine seeing matter swirl around the center of a galaxy and disappear forever into a black hole from which nothing, not even light, can escape.

Imagine an observatory that can answer questions human beings have asked since they first gazed into a night sky. Where did we come from? How did we get here? When did time begin? Are we alone in our Milky Way galaxy, with its hundreds of billions of stars, or is there life on another planet? Since there are as many galaxies in the universe as grains of sand on the beach, could life exist in those galaxies, too?

The idea of peering into the farthest and oldest corners of the known universe captures the attention of people all over the world. For years, the National Aeronautics and Space Administration (NASA) and the sixteen countries of the European Space Agency (ESA) worked together to create a telescope that would provide answers to our questions and perhaps raise some more. The Hubble Space Telescope was the astronomers' dream instrument, and its launch aboard the space shuttle *Discovery* on April 24, 1990, was heralded around the world. Once in orbit, 380 miles above the Earth, the Hubble was set free to begin its search for the outer edges of the universe.

But the astronomers' dream quickly became a nightmare. Something was wrong with the Hubble. Part of its vision was blurred. It could answer some, but not all, of the astronomers' questions. "NASA Fails!" "Hubble's in Trouble!" "Future of NASA in Jeopardy!" screamed the newspaper headlines. The public was giving up its dream, but the scientists and engineers at NASA and ESA were hanging on to theirs. They carefully constructed a plan to solve the problem. Three years later, seven astronauts worked sixteen hours a day, seven days a week, to prepare for the great adventure of fixing the Hubble Space Telescope.

Speaking shortly before he left Earth and headed into space on the repair mission, astronaut Dr. Story Musgrave said, "I'm not overconfident. I'm running scared. This thing is frightening to me. I'm looking for every kind of thing that might get out and bite us."

There were plenty of scary things that could "bite" Story and his fellow astronauts as they worked. They faced a huge problem that had to be solved. The future of NASA was at risk, and the stakes were sky-high.

CHAPTER ONE

THE HUBBLE'S TROUBLES

Astronomy is the world's oldest science. The earliest stargazers had to rely on their eyes to study the heavens. Then in 1609, Italian astronomer Galileo Galilei revolutionized astronomy with his pioneering improvements to the telescope. Galileo's new, more powerful telescope revealed the hills and valleys of the moon and the myriad stars in the Milky Way. A bold thinker, Galileo was persecuted for championing Copernicus's theory that Earth orbits the Sun, and not the other way around.

The Hubble Space Telescope has winglike panels, called solar arrays, that convert solar energy into the electricity that powers the telescope's instruments. The solar array on the right was badly twisted.

Today's telescopes "see" by collecting light from the objects they are studying and then focusing and relaying the light to instruments that analyze and record it. Most large modern telescopes use two or more mirrors to gather the light. They are called reflecting telescopes.

During the first half of the twentieth century, American astronomer Edwin Hubble worked with the great reflecting telescopes at the Mount Wilson Observatory near Pasadena, California. He made many important discoveries, including finding evidence that our universe continues to expand outward from an initial explosion of fiery gases—often called the big bang.

Like Edwin Hubble, today's astronomers and astrophysicists rely on Earth's great observatories to help them gather information about our universe. However, the telescopes in these observatories must peer through Earth's atmosphere in order to study the solar system, and that can be a problem. Our

JPL
FAIRCHILD

atmosphere is made up of a thin layer of wiggling gases that surround our planet like the fuzz on a tennis ball. The wiggle in the atmosphere causes a slight blur to appear around any object studied with a telescope. It is this wiggle that makes stars appear to twinkle in the sky when they are viewed from Earth. Stars do not twinkle if seen from space.

When flights into outer space became possible, astronomers dreamed of placing a powerful observatory in orbit, free from the blur of Earth's atmosphere. The $1.6 billion Hubble Space Telescope, named in honor of Edwin Hubble, was that special observatory.

Astronaut and astronomer Dr. Jeffrey Hoffman said about the telescope, "Astronomy has entered a new era in the space age, and Hubble Telescope is really the culmination of this. It is the premier observatory of the last part of this century and the first part of the next century."

Free from the blur of Earth's atmosphere, its instruments would see farther and more clearly than any telescope on Earth. As it explored our universe, it would gaze clearly upon the planets, comets, stars, and asteroids in our Milky Way galaxy, as well as the stars and planets in primeval galaxies.

But two months after the telescope was launched, the problems began. Like other reflecting telescopes, the Hubble uses mirrors to focus the light that it gathers from stars and planets. Like a lens in a pair of eyeglasses, the Hubble's mirror had to be precisely ground in order to focus its vision. But somebody made a mistake. The mirror was supposed to have a curve in its edges; however, one edge of the mirror was too flat by an amount equal to one-fiftieth of the thickness of the page you are reading! It does not sound like much of a mistake, but it was. Astronaut Dr. Story Musgrave said, "It was an error of common sense and an error of thinking. You didn't have to be a rocket scientist or an optician to have made that mistake."

Because the mistake was made, Hubble's instruments simply could not focus light into a sharp point. Instead, the mirror had a condition that astronomers call spherical

The Hubble Space Telescope being prepared for its 1990 launch.

aberration: the light that the mirror collected from distant objects was spread out in a kind of fuzzy halo. It had to be fixed.

And there were other problems, too. Winglike instruments called solar arrays gather energy from the sun and convert it to electricity to power the Hubble's cameras and instruments. As the telescope passed from night to day along its orbit, the arrays expanded and contracted, giving it a bad case of the jitters. The jittering was dangerous for the arrays and shook the telescope and its delicate instruments. It had to be stopped.

Then there was the problem with the gyroscopes. A gyroscope is a spinning, toplike instrument that is used to keep boats, airplanes, and space satellites steady. The Hubble had six of them. Three were used to steadily point the telescope in the direction scientists wanted to look. The other three were backups, in case a gyroscope failed. The scientists discovered that three gyroscopes had already failed, leaving no backups. If another failed, the telescope could not be pointed accurately and would therefore be useless.

The Hubble certainly was in trouble. It needed new "glasses," new solar arrays, and a set of new gyroscopes. The repair job would be expensive and enormous. Furthermore, it would have to be accomplished in outer space.

There were some people who said the repairs were so expensive, they should not be done. Others said the repairs *could* not be done. However, scientists at NASA and ESA, like most of us, did not want to be remembered for the mistakes that were made. Instead, they wanted to be remembered for how they dealt with them. The scientists felt that the Hubble could and should be repaired, so they rolled up their sleeves and went to work.

Like doctors with a sick patient, scientists worked to come up with a diagnosis and then a cure for their ailing telescope. While they planned for the major repairs ahead, astronomers made the best use of their valuable instrument by using computers to help sharpen some of the Hubble's fuzzy images.

After much thought NASA decided to build two instruments for the Hubble. One was brand-new, and the other was an improved version of an instrument that was already in the telescope. The new instrument went by the long name of Corrective Optics Space Telescope Axial Replacement, but everyone called it COSTAR for short.

COSTAR is an instrument that looks like a telephone booth and is about the same size. It contains ten small mirrors, not much larger than the ones dentists use to see inside our mouths. The mirrors have been shaped in such a way that they undo, or subtract, the error in the Hubble's primary mirror. COSTAR's mirrors are mounted on small mechanical arms. They can be retracted, like a turtle's head, to the safety of COSTAR's box, until the moment they are deployed and become the Hubble's new "glasses."

The other instrument, an improved version of the Wide Field Planetary Camera II or WF/PC II (pronounced "wiff-pic two" by everyone associated with it), was in the planning stages before the Hubble trouble began.

(Right) A scientist inspects COSTAR's tiny mirrors before they become Hubble's new "glasses."

(Far right) In Bristol, England, scientists create new solar arrays for the telescope. These help the telescope maintain an even temperature and prevent the "jitters."

SPACE SYSTEMS

FOR EVA HANDLING ONLY
FOR EVA HANDLING ONLY

(Left) Clothed in sterile garments to prevent contamination of the instrument, technicians prepare WF/PC II for its flight into space.

WF/PC II was developed to study the objects at the farthest and oldest reaches of our known universe. When the Hubble's vision problem became apparent, WF/PC II also had some corrective glasses built into it.

In addition to WF/PC II and COSTAR, engineers in Bristol, England, designed and built new solar arrays with thermal blankets to maintain an even temperature and prevent the jitters.

When the repairs were completed, astronomers said that the Hubble would be so powerful it would be like standing in Los Angeles, California, looking at the space shuttle on its pad in Cape Canaveral, Florida, and being able to count the stars on the American flag painted on the shuttle's wing!

Next NASA had to find the right people to do the job. Story Musgrave seems to believe it was his destiny to do this work. When he was asked why he gave up being a surgeon to become an astronaut, he smiled and answered quickly, "Why, so I could operate on the Hubble, of course."

(Right) During the mission, Story Musgrave peers into Endeavour *from outer space and greets the crew inside.*

CHAPTER TWO

THE ADVENTURERS

This $690 million operation, or mission, had a number instead of a name. It was called STS-61. STS stood for Space Transportation System, referring to the space shuttle (or orbiter, as NASA calls it). The crew of STS-61 was scheduled to travel into space aboard the shuttle *Endeavour*. The name seemed especially appropriate, since the word *endeavor* means "to try hard; make an effort; strive."

"The Hubble's our eyes. That's how we're going to see back to the beginning of the universe. It's how we tell what's happened to us in the past and where our universe is going in the future," said astronaut and mission specialist Kathryn Thornton, Ph.D., as she prepared to work on the Hubble during the eleven-day repair mission in space. In addition to being a veteran spacewalking astronaut, Kathy is a physicist, a scientist who studies the laws of nature and the universe. She is married to another physicist, Dr. Stephen Thornton. They are the parents of three daughters, Carol, Laura, and Susan. When Kathy isn't working as an astronaut, she says, "I spend most of my time in kids' reality."

Kathy encourages young girls interested in becoming astronauts to "study the hard sciences—math and science, engineering, physics, medicine. They are the skills NASA will be looking for, for the space station and lunar colonization."

Kathy herself studied hard sciences when she was in school, but she never

(Right) Tom Akers and Kathy Thornton celebrate their last training run in the WETF by wearing suspenders that match the identifying stripes on the legs of their space suits.

(Far right, top) Kathy Thornton communicates through the sound system wired into her "Snoopy Cap."

(Far right, center) Kathy and her daughter Susan build a toy space station.

(Far right, bottom) Kaye Akers finishes a quilt she named STS-61 in honor of her husband's mission.

considered becoming an astronaut. "It was not an option when I was a kid," she said. "There were no women astronauts, so I never much thought about it—although we used to play astronaut. We'd get a tape recorder and a couple of walkie-talkies and pretend like we were going to the moon. We'd go to the end of our driveway on our skateboards and that was our lunar landing."

Kathy is now the most experienced woman astronaut in the world, but she has said, "I would *love* to see my record broken."

Kathy's partner for the mission, the astronaut who would accompany her through the shuttle's air lock and out into space, was Lt. Col. Thomas Akers of the United States Air Force. Tom has a family, too. He is married to Kaye, and they are the parents of two teenagers, David and Jessica (or Jessi, for short).

If someone had asked Tom what he wanted to be when he grew up, chances are he would not have answered "an astronaut." While he was in college studying applied mathematics, Tom worked as a forest ranger. When he graduated, he returned to his and Kaye's hometown of Eminence, Missouri, and became principal of Eminence High School—his old alma mater. While still a high school principal, Tom said, "I went out looking for an easier job." He smiled, then added, "I didn't become interested in flying until I was about thirty, when I joined the Air Force and became a flight-test engineer. Being an astronaut is the ultimate job for a flight-test engineer. Getting to go into space is a great job."

Kathy and Tom were joined on this mission by five other astronauts, who also had families, homes, and private lives that would have to be put on hold as they worked and trained together for nearly two years in order to be ready to fix the ailing Hubble.

Air Force Col. Richard Covey was the mission commander. Dick was to fly the shuttle and have overall responsibility for the mission's success and the flight's safety. As this quiet man talked about the mission he said simply, "Our task is to find the Hubble and fix it. To do that we need good people, good preparation,

Astronaut Dick Covey, who occupied the commander's station aboard the Endeavour. *Although Dick is a graduate of the U.S. Air Force Academy and Purdue University, many of his trainers attended Texas A&M University, and he wore an Aggie cap in their honor.*

and good fortune."

Dick was backed up at the shuttle's controls by pilot Kenneth Bowersox, a commander in the United States Navy. Sox, as he is called, had dreamed of being an astronaut since he was six years old. "I heard stories about John Glenn going around the Earth and I thought that would be really neat. And so as I went on through school, I kept that as a goal. It's been a wonderful experience to get to fly in space, and I can't wait to do it again."

Another set of astronauts who couldn't wait to do it again were Dr. Story Musgrave and Dr. Jeffrey Hoffman. They were the second team of spacewalking astronauts for this mission. Although the nickname irked Story, he and Jeff were called the Odd Couple because they did their spacewalks on the odd-numbered days, whereas Kathy and Tom—the Even Couple—worked on even-numbered days. In addition to being a surgeon, Story holds university degrees in mathematics, business administration, chemistry, physiology, literature, and philosophy! Story was the payload commander for this

(Far left) Ken Bowersox wears a partial pressure suit during the launch and reentry phases of the flight. If, during launch, an emergency evacuation of the shuttle is needed, the suits protect the crew from exposure. During reentry, the suit maintains pressure on the lower part of the astronauts' bodies. This prevents blood from pooling there and causing blackouts.

(Left) Once in orbit, Sox can pilot the shuttle wearing comfortable clothing.

mission, which meant that he planned and coordinated the "house calls"—space walks—the astronauts took as they operated on the Hubble. Story said, "I'm here because I love space, I love being in space, and I like the space business. You can't separate out space and the space business. I've been an astronaut for 26 years, and I've had roughly 26 days in space. I've got only 1 day in space for every 365 days down here. It's not a very good ratio, but still I love space so much that I would continue to do it, even if the ratio were worse."

Dr. Jeffrey Hoffman is an astronomer and astrophysicist who was

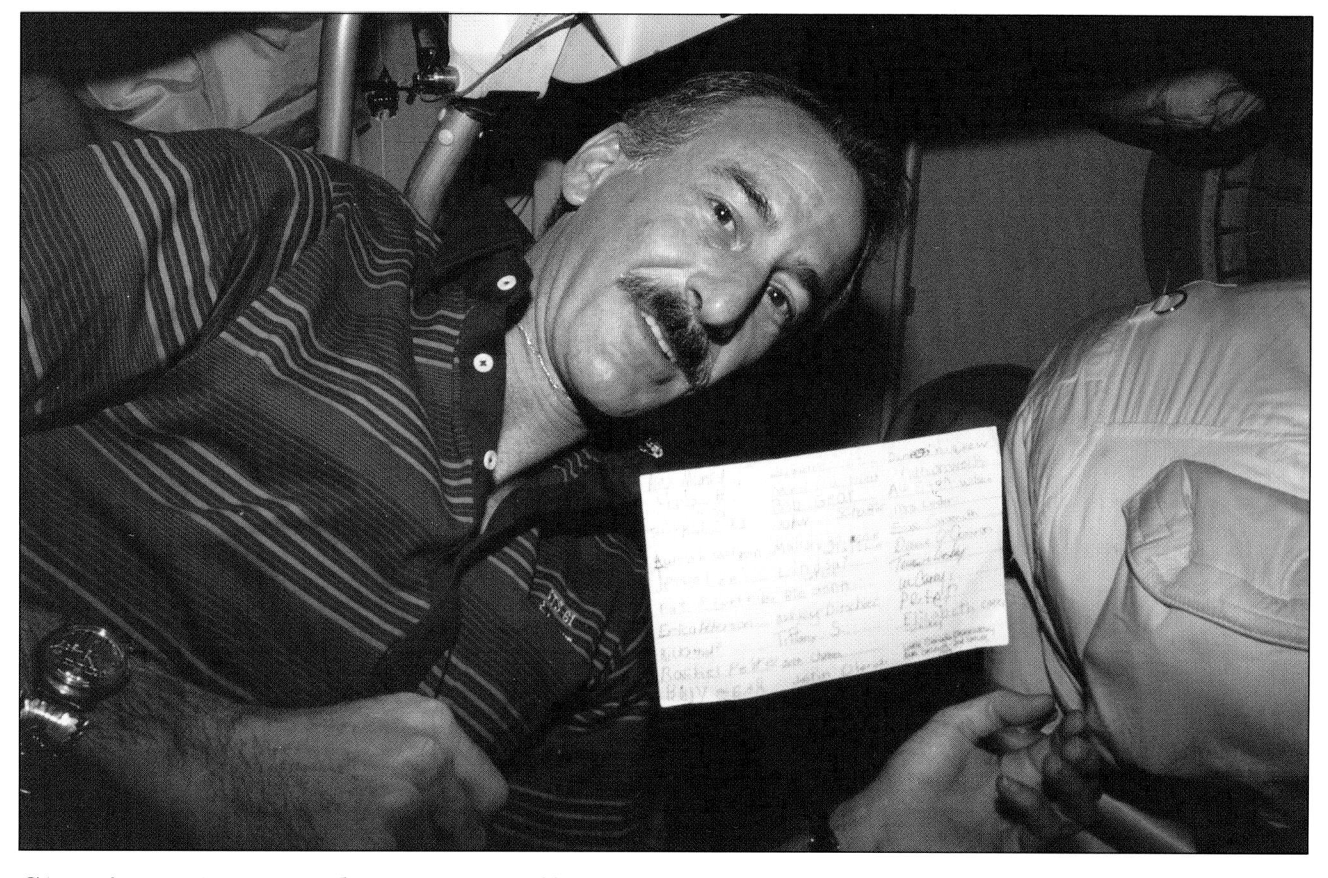

Elementary school children signed a card and sent it into space with Jeff Hoffman.

Story's partner on the space walks. Asked about his career as an astronaut, Jeff said, "I think the lure of space has been drawing me since I was a little kid, six years old. Back then, there was no such thing as an astronaut. I used to go to the planetarium. Astronomy fascinated me because that involved space, something about the unknown, exploration, learning things that people didn't know before or doing things that people hadn't done before. I feel fortunate to be alive when we're doing this sort of thing."

Astronaut Claude Nicollier represented the European Space Agency on this crew. Claude is an astro-

Story Musgrave gets some help putting on his extravehicular visor assembly (EV/VA)—or space helmet—in preparation for a training run on the air-bearing floor. Because of its weight, the top portion of the space suit was supported on a rack and Story slipped up into it and lowered himself out of it. Life-support systems are housed in the backpack. Most space suits last about eight years before they are replaced.

physicist and a captain in the Swiss Air Force. "This mission is important for Europe because of our overall participation in the Hubble Space Telescope program," he said. "There are a lot of people in Europe who have this telescope in their hearts, and we want to serve them well."

Claude's job on this mission was operating the fifty-foot robot arm that would carry a spacewalking astronaut up and down the telescope and all around the shuttle's payload bay. Speaking of Claude, Kathy Thornton said, "Claude can make that arm go anywhere. It can almost

Working on Endeavour*'s flight deck, Claude Nicollier moves the astronauts using the Remote Manipulator System (RMS), or robot arm. The controls for the RMS are left of center in this picture. A team of spacewalking astronauts can be seen through the shuttle's window.*

wrap itself around the space telescope in the payload bay. Claude gets us where we need to go. He almost knows before we tell him."

This, then, was the team of adventurers who put their brains, their skills, their dedication, even their lives on the line in order to repair the greatest telescope in the world. The training was grueling, the hours long. There was much to study, much to learn. Yet Tom Akers said jokingly, "I watched my tires in the parking lot every day because I was sure they'd be slashed. Everyone wanted to be assigned to this mission!"

This team of astronauts was the most experienced crew NASA had ever assembled for a flight. There were no rookies, or first-timers, among them. NASA's reputation, for good or ill, traveled with them on this mission.

CHAPTER THREE

PRACTICE, PRACTICE, PRACTICE

There is an old military expression that goes, "Proper prior planning prevents poor performance." Proper prior planning . . . that is exactly what everyone at NASA did as they prepared for this mission.

Talking about the job ahead of her, Kathy Thornton said, "It's not so important how fast we get the job done, but it's very important that we do it right and very carefully." That is a good rule to apply for any of us who have a problem to solve. Kathy knew that fast is not always best. There would be no room for second chances once the astronauts left the shuttle and began to work in the weightless environment of space.

And Jeff Hoffman said, "Although we are dealing with a very complex scientific instrument, our job as astronauts is not an intellectual task. Here with the Hubble Telescope we're technicians, we're mechanics. We're actually going to go up there and use our hands to fix things."

Tom Akers added, "The ground team has come up with the plan, and the astronauts go out and execute it. We don't have a better computer in the world than the human brain, when it comes to solving problems."

Story Musgrave had a final word. "Running scared keeps you thinking about how to get the job done."

So how *were* the astronauts going to get the job done? By planning ahead, paying attention to the details of each task, and practice, practice, practice.

The astronauts were going to work in teams on the space walks, or extra-

Wearing training versions of their space suits, Story Musgrave and Jeff Hoffman logged hours of practice in the WETF facility at Johnson Space Center. Here they practice sliding an exact replica of the WF/PC II into a mock-up of the Hubble Space Telescope.

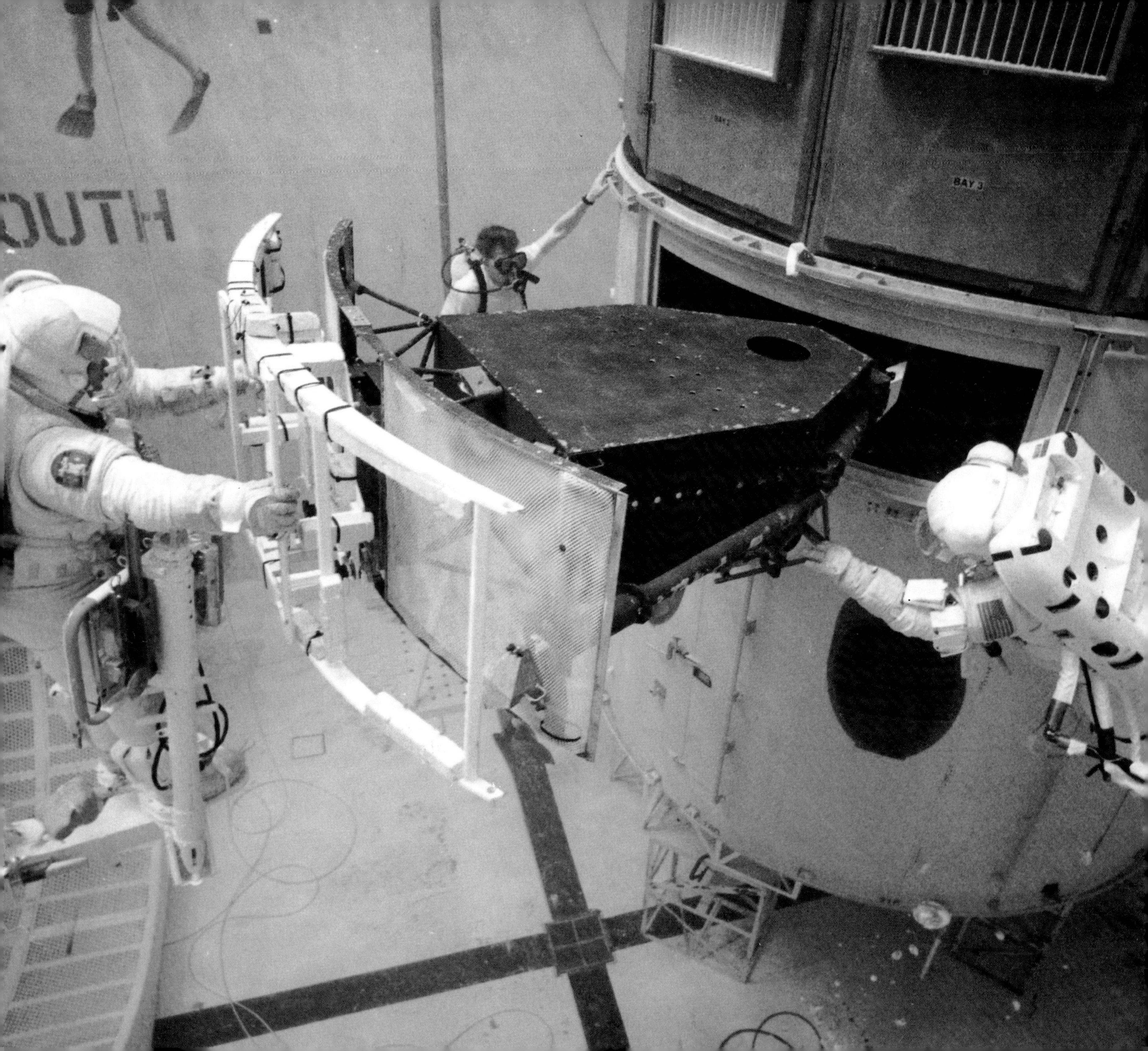
OUTH
BAY 3

vehicular activities (EVAs, for short). They would remove the damaged instruments from the Hubble and insert the new ones. One astronaut would float freely in the shuttle's payload bay while the other would ride on the end of the shuttle's long robot arm.

"Claude will move us around like a repairman in a cherry picker," Kathy Thornton said. "You have to have your feet tied down, so you can use both hands to work." There was a movable foot restraint that attached to the shuttle's robot arm as well as to other places in the payload bay.

In order to train himself to move the arm with extraordinary precision, Claude spent hours working with a training version of it—complete with a dummy astronaut on its end! Ken Bowersox trained with the robot arm as well. All of the astronauts were cross-trained; that is, each of them was trained to do someone else's job, in case of an emergency.

The astronauts rehearsed every single move they would make during their repair mission. To help them come up with a perfect performance, they used many training aids, includ-

(Left) To get ready for the mission, Claude Nicollier practiced with a training version of the robot arm. The "astronaut" on the end of this arm is actually a mannequin.

(Right) Jeff Hoffman (wearing the hood) and the rest of the astronauts utilize virtual reality technology for part of their training.

ing virtual reality, a thirty-five-foot-deep pool called the Weightless Environment Training Facility (WETF, pronounced "wet-eff"), an air-bearing floor, and thermal vacuum chambers. The astronauts and the ground crew conducted six joint integrated simulations (JISes), which were full mission simulations—a kind of dress rehearsal for the flight itself, from launch to landing. No other crew trained as hard or as long as this one. There was simply no room for mistakes.

Visitors to the Johnson Space Center often ask to see the room where astronauts practice floating in zero gravity. The truth is, there is no such room because long periods of zero gravity cannot be duplicated on Earth. Instead, the astronauts simulated the experience of working in a weightless environment. In order to prepare themselves for becoming weightless and moving the weightless mass of the COSTAR and WF/PC II, the astronauts wore training versions of their space suits and went to work.

COSTAR and the WF/PC II were both heavy, massive objects. On Earth each weighed about six hundred pounds. In space these objects would become weightless, just like the astronauts. However, they would still take up the same amount of space, or volume, that they did on Earth. They would retain their mass; that is, although they would weigh nothing, it would still be difficult to start them moving or to stop them once they were in motion. It was the mass of COSTAR and WF/PC II that the astronauts had to practice moving on Earth before they attempted doing it "upstairs," as Story Musgrave liked to call outer space.

Mock-ups of the Hubble Space Telescope, the WF/PC II, the solar arrays, COSTAR, the robot arm, the shuttle payload bay, and all of the astronauts' special tools were ready for their training runs. When it was time for a run in the WETF, the engineers who designed the tools and the equipment came to watch. An ambulance pulled up, ready to carry an astronaut or a diver off to get help in case of an accident. Safety was tops on everyone's list.

The astronauts got ready to put on their space suits. First came either an adult diaper or a urine collection device. Most space walks (and training runs) last at least six hours, and there is no returning to the shuttle to go to the bathroom! Next was the liquid cooling and ventilation garment, a one-piece suit that looked like a pair of footed pajamas with tubes running through it. The tubes carry water to cool the astronaut during heavy work periods.

Then, with the help of technicians, the astronauts got dressed in their space suits, which are called extravehicular mobility units, or EMUs, for short. Tom had to lie down to get into what we would call the pants (but NASA calls the lower torso assembly). Boots were next, then the hard upper torso (or HUT). Built into the HUT was a backpack called a primary life-support system (PLSS, pronounced "pliss") that contained all of the life-support systems needed in outer space. There was

(Top right) On Earth and in space, astronauts need help donning their space suits. Tom Akers lies down to put on the lower torso assembly (LTA) of his suit.

(Bottom right) This LTA sits and waits for its occupant.

(Far right) A space suit weighs approximately 250 pounds, so astronauts cannot wear them on Earth without support. Kathy Thornton is held up by brackets as she waits to be lowered into the WETF.

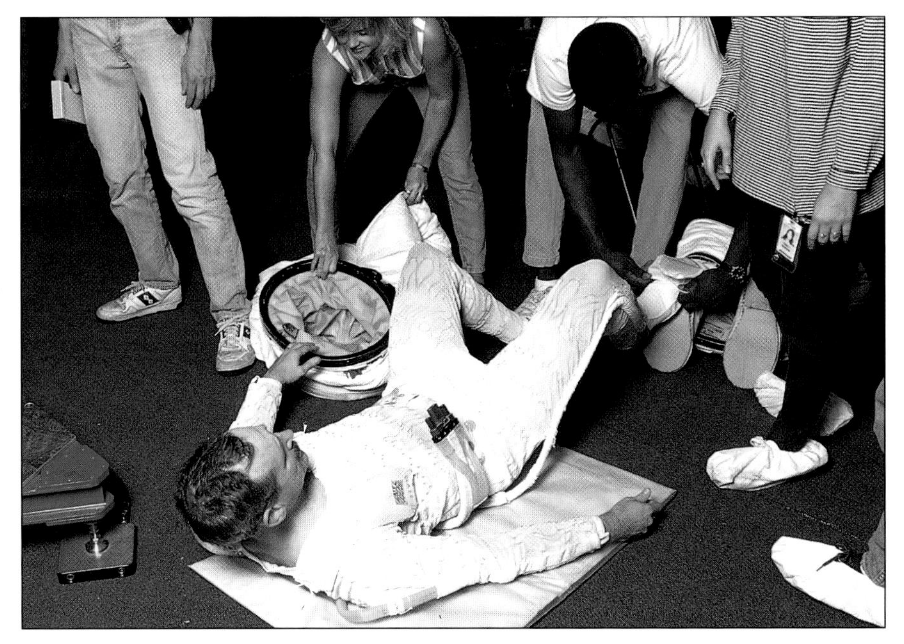

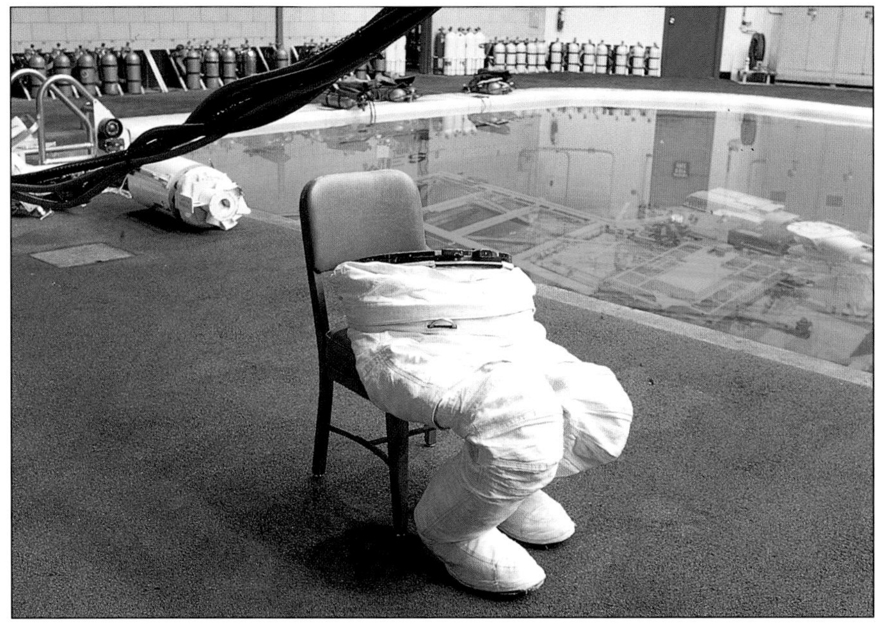

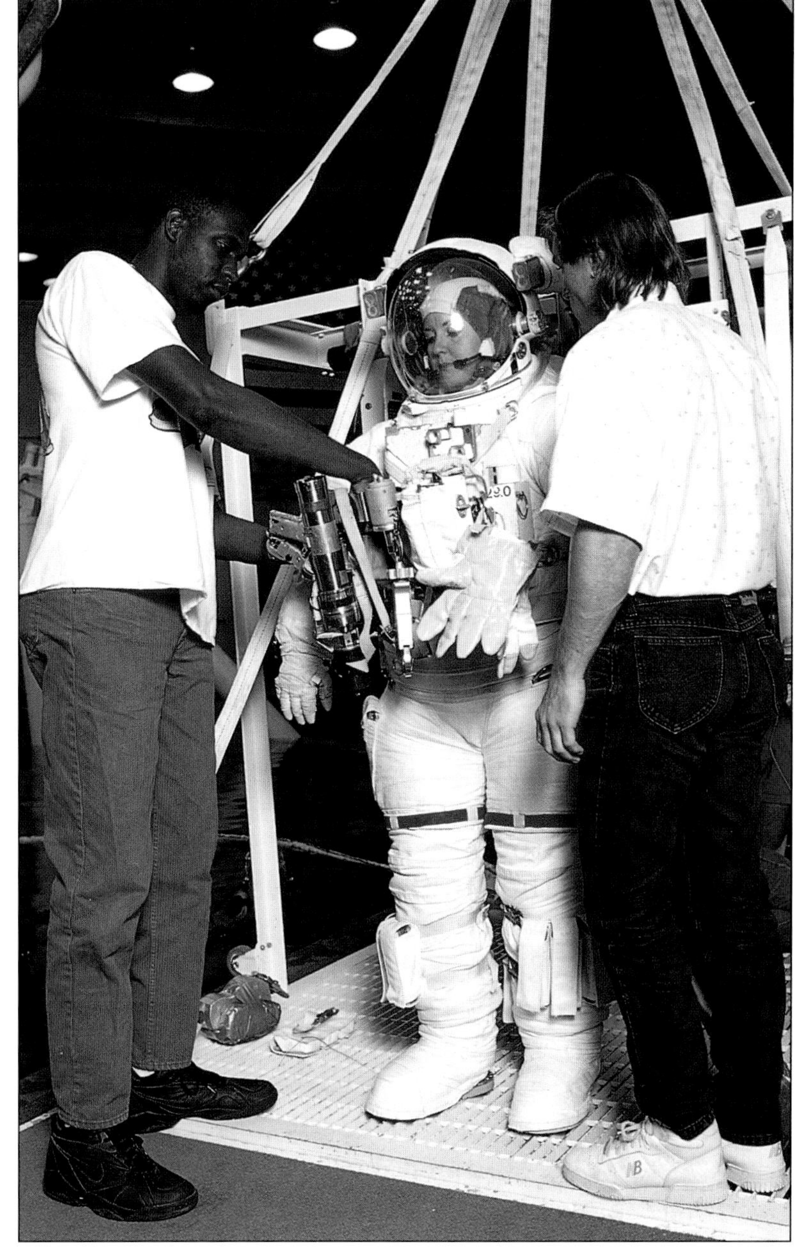

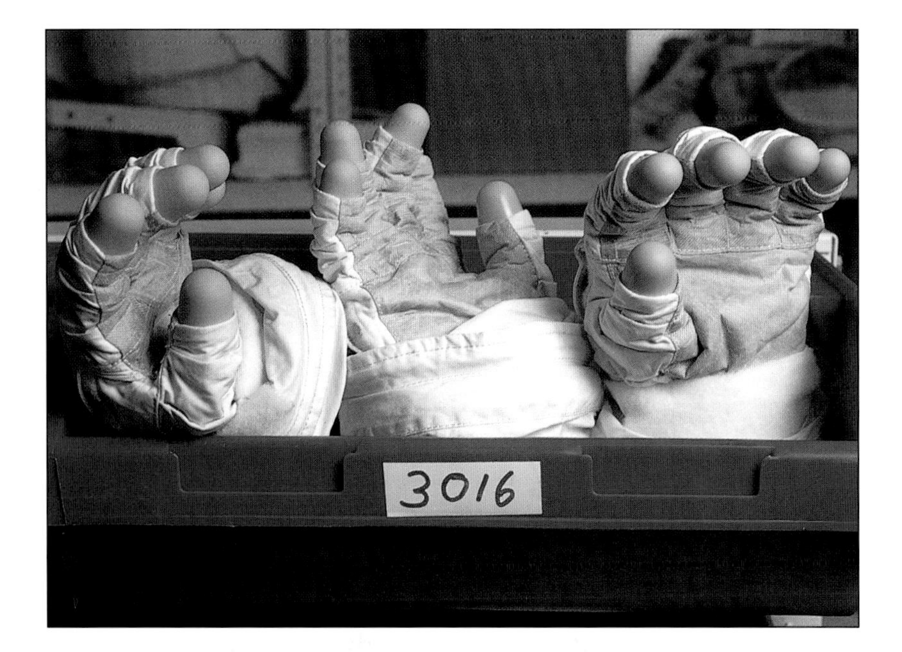
3016

(Far left, top) Gloves are the only element of the space suits that are custom-made. The other parts of the suits are adjusted to accommodate each astronaut.

(Far left, bottom) Before and after each training run, all parts of the space suits are carefully inspected and stored.

(Left) Kathy Thornton is slowly lowered into the WETF. Divers waiting below will check the suit's systems, then release Kathy from the crane to begin her training run.

oxygen for breathing and pressurizing the space suit, filters for removing carbon dioxide, and a system to cool and circulate the water in the ventilation garment. The HUT was too heavy to be worn by an astronaut who still felt gravity, so it was attached to a support system, called a donning station because that was where Tom and the rest of the astronauts donned, or put on, the suit. Astronauts slipped up into the top of the suit and technicians put the two pieces together. A soft fabric cap called a "Snoopy Cap" (because it resembled the one the famous dog wears when he is pretending to be a World War I flying ace) was put on. The Snoopy Cap had headphones and microphones for two-way communication and was equipped so the astronaut could hear caution and warning tones. Next came the gloves—the only part of the space suits that are custom-fitted for each astronaut—and finally, the helmet. Once the helmet was in place, the astronaut began to breathe from the air in the PLSS. There was an additional supply of oxygen attached to the PLSS, in case the first supply failed. The oxygen in a space suit lasts for about seven hours. Space suits also protect astronauts from micrometeoroids, radiation, and ultraviolet light.

It's hard to tell astronauts apart when they are working in space, so each suit was marked differently. For this mission, Kathy Thornton's suit had a straight, broken red stripe on the legs. Tom's broken red stripe ran diagonally. Jeff had a solid red stripe, and Story had no stripe at all.

Since training runs and space walks last for several hours, the astronauts were likely to get hungry or thirsty. Inside each suit, near the neck, was a drink bag with a straw and a fruit roll-up bar that could be bitten off and chewed, a bite at a time. Hands are no help for eating in a space suit!

When the astronauts were finally dressed, technicians checked out all of the suits' parts on a checklist, just as a pilot checks things off before takeoff. Once everyone was satisfied that the suits were properly pressurized and were not leaking, the astronauts were ready to be lifted off the donning station and lowered into the water.

A model of the Hubble Space Telescope waited in the water tank. Slowly and patiently, the astronauts practiced working together to slide the instruments into their proper places on the observatory. An astronaut on the end of the robot arm moved the instrument while another astronaut—the “free-floater”—coached.

“Easy, K. T., easy,” Tom Akers said, using the crew’s nickname for Kathy Thornton.

The good news was that COSTAR and WF/PC II had been designed to slide in and out of the Hubble on rails, much like a drawer slides in and out of a cabinet. The bad news was that, as with a drawer, there was very little clearance at the top and bottom. The astronaut had to be certain that the instrument was sliding in properly, or it could jam. The worst news was that the astronaut doing the sliding—in this case, Kathy—could not see what she was doing. “It was easy to move, but all I could see in front of me was a box,” she said. Tom had to become her eyes. Fortunately, Kathy and Tom had worked together as an EVA, or

Kathy Thornton and Tom Akers spent hundreds of hours practicing in the WETF with a full-scale mock-up of the Hubble Space Telescope.

spacewalking, team before. They knew each other well and could anticipate each other's moves. Tom said, "Teamwork in EVA is very important. Communication is very important. We feel lucky to be working together again."

Being weightless in water is not *exactly* like being weightless in space. Astronauts talk about the water's viscosity—its thickness. To push something like COSTAR underwater takes some effort because the thickness of the water pushes against you as you move through it. It slows you down. On the other hand, space is a vacuum. There is nothing in it—no air, no water, nothing—to resist your efforts as you move there.

Kathy compared moving objects in space to training in the WETF. "Heavy objects sink to the bottom of the pool, and you think 'It won't happen in space.' But the water also helps you. You have to think about that and say, 'Now, could I have really moved like this in the absence of this water? Could I stop myself in space?'

"Newton's laws are in effect whether you are in orbit or not," Kathy continued. She was talking about Newton's first law of motion, which says that objects at rest tend to stay at rest, and objects in motion tend to stay in motion.

Working in the water tank was only one of the ways Kathy, Tom, Story, and Jeff trained for the Hubble repair. The WETF helped them get used to weightlessness, but it didn't help them practice moving massive objects in the vacuum of space, where there would be no resistance to their efforts. For that training, they turned to the air-bearing floor.

The air-bearing floor was first used to rehearse for the Apollo moon program. The astronaut is suspended above the floor in a harness. Pads beneath him shoot jets of air onto the smooth and level metal floor, creating an almost frictionless environment. It's a bit like practicing on an air hockey table! The astronauts can move back and forth, with no drag, giving them a freedom of movement that is impossible in the water tank.

There is another aspect of working in space that is far different from working on Earth—the temperature. It can rise to 250 degrees Fahrenheit and fall to negative 250 degrees Fahrenheit. To be certain their tools will not break in the extreme temperatures of outer space, astronauts test them in thermal vacuum chambers that duplicate those conditions.

While Story Musgrave was testing some tools for this flight, he suffered a mild case of frostbite to his hands. The chamber was cooled to 130 degrees below zero. Story's hands did not suffer permanent damage, but after that incident NASA added another layer of insulation to everyone's space gloves.

The frostbite that Story got in the vacuum chamber was a real surprise—one of those "bites" that he talked about. Tom Akers said, "We don't want surprises," but the trainers deliberately provided some anyway—just to help the astronauts practice. The trainers might fix a bolt so it wouldn't slip into place as it should. Maybe an instrument wouldn't slide properly on its rails. What would the astronauts do? Over and over the astronauts practiced, thinking about every possible thing

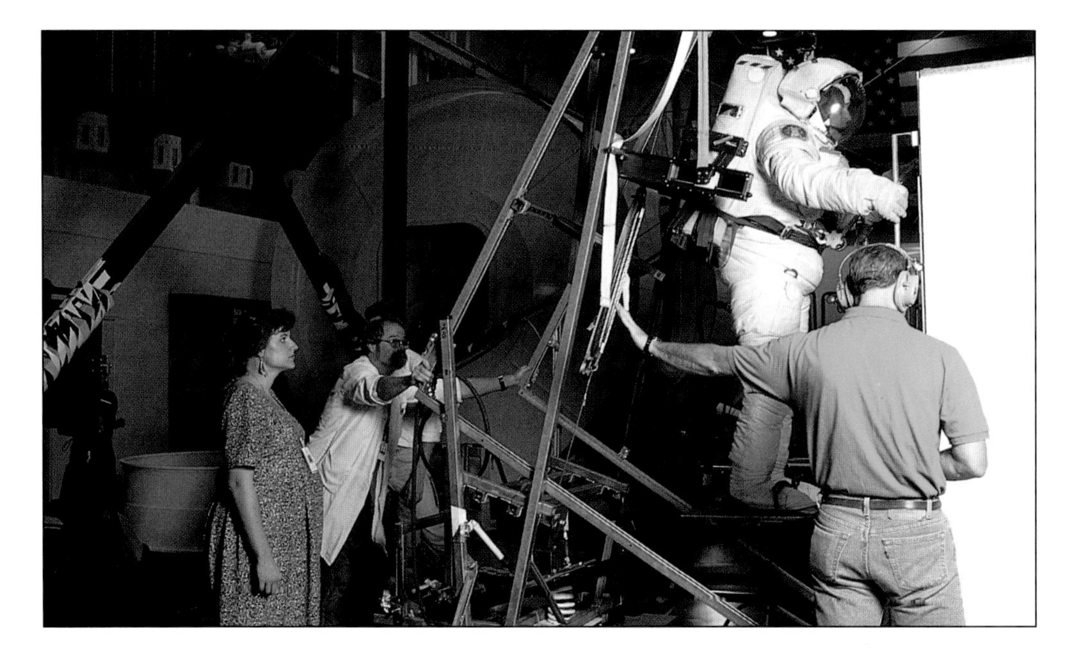

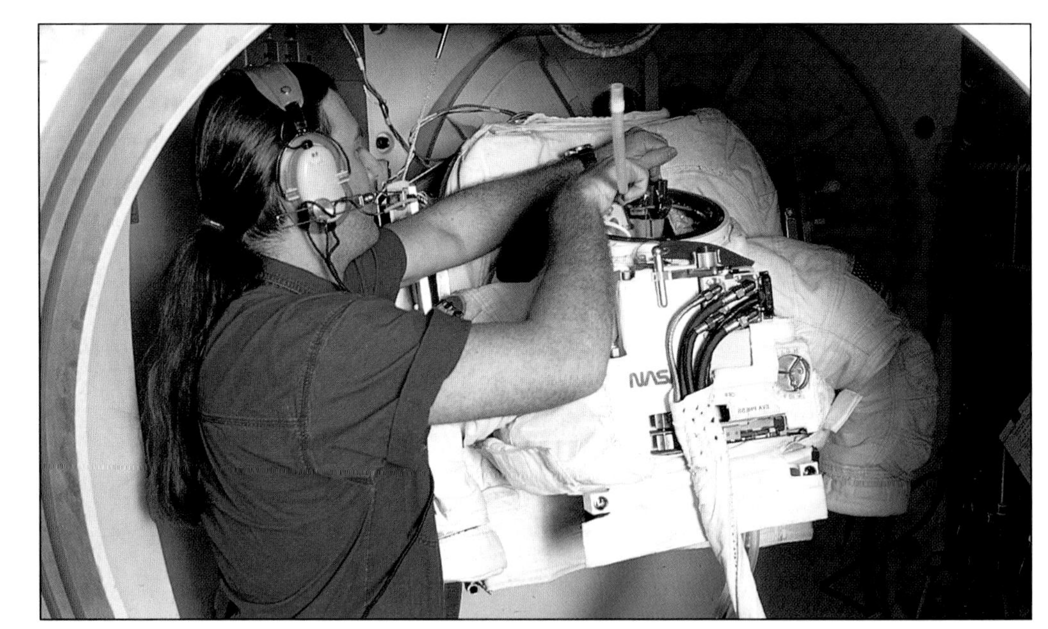

(Top) NASA engineers watch carefully as Tom Akers coaches Story Musgrave during a training run on the air-bearing floor. All of the astronauts cross-trained with each other, although they would each work with only one other person during their missions.

(Bottom) A technician prepares an upper torso assembly (UTA) before testing it in the thermal vacuum chamber.

that could go wrong and planning what they would do in case it did.

Plans are important to problem solving, but so is flexibility and the willingness and ability to change a plan if necessary. As Tom Akers said, "Not only do you need to be educated and understand the problem, but you must look at all the what-ifs before you get there. During a problem situation, that kind of planning will always pay off."

At last, the long hours, days, weeks, months of training were finally over. Story said the crew had synthesized all of their training experiences: "We've mentally built how it's going to be upstairs." But there still could be surprises. Sox said, "The biggest thing I'm worried about is the unknown factor—things we haven't thought of or things we haven't learned about. Those kinds of things are always out there. I think they've been called the 'unknown unknowns.' If we knew about them, we'd be able to fix them."

Unknown unknowns—those are the things that can "bite" you.

Story Musgrave eagerly anticipated this mission. Before the flight he admitted to a dream—he hoped to make contact with an extraterrestrial! "The greatest thing that could happen to me would be for creatures to come from space and pick me up and take me away. I know there are people who claim they've been on rides in foreign vehicles, but this is not one of those things. This is a matter of possibilities. It's almost a statistical certainty that there are beings out there millions of years more advanced than we are. The possibilities are absolutely immense. It's one chance in a billion I'd succeed in making contact, but what have I got to lose? And anyway, it's fun to try."

The training was over, and the adventure was about to begin. Many people have jobs that require them to travel far away and say good-bye to family and friends. The astronauts had to do that, too, but these adventurers were headed for the most exotic destination of all.

CHAPTER FOUR

BLASTOFF!

Jessi Akers's father, Tom, has flown into space twice. "The first time I didn't want him to go," Jessi said, "and I cried. The second time, I worried a little. This time should be okay. I wouldn't miss a launch." Her mother, Kaye, said that when she watches her husband blast into orbit, "I feel every emotion you can have, all at once. I'm happy, scared, and proud."

Kathy Thornton's daughters were worried about the danger, but Kathy tried to reassure them. "I tell the kids that *I* get scared watching them go off on a bicycle because something could happen to them," Kathy said. "But I have to let them do it—it's part of growing up. I can't stop them because it scares me, because there's some risk involved. You weigh the risks and the benefits. If it's worth it, you go for it." Kathy stopped and smiled. "I'm not doing bungee jumping or anything like that!"

The children's last hugs and kisses had to be given seven days before the launch. That was when the astronauts went into a kind of quarantine called "health stabilization." They moved out of their homes and into special living quarters at the Johnson Space Center. Husbands, wives, and essential NASA employees had to have physical examinations and be pronounced healthy before they could visit the crew during this time. No one wanted to give a case of the sniffles to an astronaut. A runny nose doesn't drain in zero gravity! Since children seem to have the

At last! Before dawn, the crew of STS-61 heads for launchpad 39B and the space shuttle Endeavour.

Environmental
Stabilization Area
No
Smoking
NASA
esa
STORY

(Left) Weeks before the launch, Endeavour *was moved to the Vehicle Assembly Building at the Kennedy Space Center, where it was attached to its external fuel tank and twin solid rocket boosters.*

(Right) From prelaunch activity to reentry into Earth's atmosphere, every moment of the flight was carefully monitored at Mission Control.

most runny noses of all, the astronauts' children could talk to them on the phone, but they were not allowed to be with their parents during this period.

During the quarantine period for STS-61, the trainers began to adjust the astronauts' eating and sleeping schedules. They wanted to move the astronauts' body clocks back in time so they would be wide awake for the launch, which was scheduled for just before five o'clock in the morning.

Kaye Akers talked about visiting Tom in quarantine and said, "I went by to see Tom on the day before Thanksgiving. It was about 9:00 A.M., and they were having their evening meal—salmon." She paused while she contemplated salmon in the morning. Then she added, "I just had a cup of coffee."

The astronauts missed having Thanksgiving dinner with their families, though they celebrated together ahead of time. Kathy Thornton said, "I tried very hard to convince the kids that it was a recently discovered historical fact that Thanksgiving was on Tuesday!"

And so the long months of planning and practice finally were over. The crew flew to the Kennedy Space Center in Cape Canaveral, Florida. In a matter of hours the flight of the space shuttle *Endeavour* was scheduled to begin. Blastoff from the launch pad was planned for 4:57 A.M. on December 1, 1993.

By launch day, press and dignitaries from all over the world had arrived at the cape. Launch morning began clear but soon turned cloudy—an ominous sign. *Endeavour* would not be allowed to leave the launch pad if weather conditions were poor.

The Kennedy Space Center is a bird and nature sanctuary, part of the Merritt Island National Wildlife Refuge. The grounds are full of pigs, deer, armadillos, eagles, and the occasional alligator. Since the public

Endeavour, *ready and waiting for blastoff. The "building" to the left, called a gantry, contains an elevator, which the astronauts ride to the top of the shuttle. They enter the spacecraft through a passageway, or hatch, near the windows. After the launch, the white solid rocket boosters on either side fall into the sea and are recovered and reused. The large red liquid fuel tank is jettisoned in space.*

would be allowed to enter the site and view the launch from specified areas, the United States Fish and Wildlife Service began a sweep for any alligators that might decide to grab a spot for themselves. A big one was found sunning itself near a viewing stand and was carried away to a safer location.

The astronauts went to bed at noon. They woke up at 8:00 P.M., ate breakfast, and began the process of getting ready for the flight. By 1:00 A.M. the press had passed through a barrage of explosives-sniffing dogs and gathered outside the building where the crew was preparing to depart for the launch pad. Kennedy Space Center SWAT teams patrolled the tops of nearby buildings. Helicopters buzzed overhead, and more teams patrolled the ground area. No one wanted to take a chance on *anything* going wrong at this point.

The astronauts appeared! Cameras flashed, a cheer went up, and reporters shouted, "Good luck! . . . Good-bye!" The van that took them to the launch pad quickly pulled away, and they were gone.

Crowds had gathered at the viewing sites. Some had brought their own telescopes, the better to see the fiery launch. While they waited, they studied the heavens.

"There's Orion," someone said. "I can see his belt."

"I see the Big Dipper!" another amateur astronomer announced.

Then there was an exited shout. "There's the Hubble!" Sure enough, the satellite telescope was streaking overhead. "Come and get me; come and fix me," it seemed to say.

The countdown continued, but the weather reports were not good. The winds were too high for launch, but launch was still a few hours off. Maybe they would die down. The rockets carrying *Endeavour* could roar through any winds on takeoff, but the shuttle lands as a glider, with no power. If there were an emergency and the shuttle had to return quickly to the Kennedy Space Center, high winds could jeopardize the landing. Safety was on everyone's mind.

Time ticked by. The radio crackled with the news that an intruder, a large cargo ship, was in the area where the shuttle's booster rockets would fall back to the sea and be retrieved. *Endeavour* could not be launched as long as the ship was there, and efforts to contact it were failing. The wind continued to blow. At T minus a few minutes, the decision was made to scrub, or cancel, the launch. Everyone, the crew most of all, was disappointed. It felt as if someone had just canceled the Super Bowl right before the kickoff.

The astronauts returned to their temporary living quarters. They would try again at 4:26 A.M. the next day, Thursday, December 2: Laura Thornton's eighth birthday.

In the early hours of that day Laura; her older sister, Carol; her younger sister, Susan; and their father, Steve, were together with the other astronauts' families on the roof of the Mission Control building at the Kennedy Space Center, where they would watch the launch. The winds were calm and the sky was perfectly clear. Everyone knew it was a go for launch. Kathleen Covey, Dick's wife, had brought Laura a birthday cake, in anticipation of a postlaunch celebration. Steve Thornton gathered his daughters around him as the countdown proceeded.

At last, the moment everyone was waiting for had come. "Five . . . four . . . three . . . two . . . one . . . BLASTOFF!" The rocket engines ignited, the dark sky exploded into a fiery light from the shuttle's two solid-rocket boosters, and the 4,511,115 pounds that made up *Endeavour* and its payload slowly rose from the launchpad, gathered speed, and streaked like a bright orange fireball across the dark sky.

As *Endeavour* climbed, Commander Covey's voice came over the radio. "It's a beautiful sunrise," he said. The spacecraft was gradually becoming smaller and smaller. Now it was the size of a star . . . now it was disappearing into the darkness, heading toward its rendezvous with the Hubble. NASA's do-or-die mission was on its way.

A voice came crackling over the radio while the sound of the rocket's roar still hung over the space center. It was Kathy Thornton. "Tell Laura how'd she like *those* candles for her birthday!"

(Left) Laura Thornton (seated, right) celebrated her ninth birthday many times! After her mother's return from space, Laura shares cake and ice cream with her parents and sisters Susan (lower left) and Carol (standing).

(Right) A fiery blastoff into the still-dark skies over Cape Canaveral.

USA
NASA
Endeavour

CHAPTER FIVE

JUST FLOATING AROUND

When *Endeavour* began chasing the Hubble, it was 4,600 miles behind it. Each time *Endeavour* orbited Earth, the shuttle gained 375 miles on the telescope. Since an orbit takes only one and a half hours, it took two days to close the distance. The astronauts spent that time checking out all of the shuttle's systems and getting used to being weightless, a sensation everyone seems to love. Perhaps Sox put it best: "Floating around? That can't be beat for pure fun."

The shuttle's robot arm (lower right) is poised, waiting for the right moment to capture the Hubble Space Telescope and anchor it in the orbiter's payload bay.

So what makes an astronaut weightless in space? In order to understand what happens, you must know something about gravity, microgravity, and force.

Way back in the seventeenth century, Isaac Newton watched an apple fall from a tree. Newton had a wonderful imagination, and he wasn't afraid to use it. He reasoned that the force—gravity—that pulled the apple down to the ground also extended out into space to pull on the moon as well. Then he began thinking about our spherical Earth. He pictured the tallest mountain in the world with a cannon on top of it. He thought about the cannon firing a cannonball. He knew that the ball would travel more or less parallel to the ground for a while, then gravity would gradually pull it back down to Earth in a curving line called an arc.

Newton thought about the two forces that were at work here—thrust, the force that moved the cannonball out, and gravity, the force that pulled it down.

If the cannon had enough force to send the cannonball along a parallel path way out from Earth's surface, gravity would still work to pull it down. But in this case the path would not be an arc. Instead, gravity would pull the cannonball in a complete circle around the Earth, and it would return to where it started—something we now call an orbit.

Today's rocket scientists use Newton's theories to put the space shuttles in orbit. The rockets and solid-rocket boosters push the shuttle out from Earth with force and speed. *Endeavour* entered orbit at a speed of 17,500 miles per hour. Then gravity—not as much as we have here on Earth, but gravity just the same—took over and bent the shuttle's path into a circle that matched the curvature of Earth. The shuttle fell around the Earth in a circle, and so did everyone and everything in it.

Which brings us to why astronauts float in space. The fact is, the astronauts, the shuttle, and everything in it are falling rather than floating. But they are all falling together at the same speed. However, the astronauts do not feel as if they are

(Left) A weightless Kathy Thornton. Like those of all the astronauts, her face appears slightly puffy. While weightless, body fluids are evenly distributed because there is no gravity to pull them downward.

(Right) Space mechanic Jeff Hoffman displays his tools and tool belt in the shuttle's mid-deck. The temperature inside the crew compartment can be regulated from 61 to 90 degrees Fahrenheit, so the astronauts are able to work in comfortable clothes.

falling as they float around suspended inside the shuttle. Believe it or not, you have experienced the sensation of free fall, weightlessness, zero G, or microgravity (different names for the same experience) if you have ever ridden a swing high and hard. As the swing comes to the top of its arc, you feel the chains go slack, and for an instant you lift up off the seat. During that instant you feel weightless because you are in free fall. As soon as that moment passes, gravity pushes you back into your seat for the ride back down.

Since the crew compartments are pressurized and temperature-controlled, each astronaut can wear the same kind of clothing he or she would wear back on Earth as they float around inside the cabin. But NASA provides the astronauts' clothing—right down to their socks and underwear! NASA controls the clothing because it must be treated before it can go on a flight. Every

(Left) Claude Nicollier and Dick Covey take sponge baths aboard Endeavour. *Clean water comes from a water pistol, and dirty water is collected with a sponge and squeezed into the shuttle's waste collection system.*

(Top right) Liquids that float freely in space form into round balls, so drinks are contained in foil packages and sipped with a straw. Here Ken Bowersox recaptures a ball of juice that "escaped" from his straw.

(Bottom right) Everything the astronauts eat and drink is carefully tested and evaluated before being approved for spaceflight. A panel of engineers, dietitians, and food specialists samples a selection of juices at the Johnson Space Center.

piece of cloth that goes into the shuttle is washed and dried at least thirty-six times to remove all traces of lint. Lint can clog air filters and fog up windows. There is no way to get rid of it once it's there, so the idea is to take as little of the pesky stuff along as possible.

The astronauts' food is far better than it was in the early days of spaceflight, when goop was squeezed out of a tube or eaten in gelatin-coated cubes. Now their meals are like what you eat at home. However, most of the food is precooked and processed so it requires no refrigeration. Fresh fruits and vegetables travel to space, but they must be eaten during the first two days of flight, before they spoil. There is no room for a table, so astronauts eat from a food tray that is attached to their clothing by a Velcro strap. Some hang from the ceiling to eat, some eat strapped to a wall, and others just float and chew.

Sleeping in weightlessness can be a problem. Astronauts sleep in special sleeping bags that have a firm support for their backs. "Down" is wherever their back is.

Story Musgrave has always slept on his side here on Earth. His body was used to that position. As Story put it, it was as if his body were saying, "If you don't put me on my side, you're going to have a tough night." Story continued, "I try to tell it, 'It won't you do any good in space,' but my body wants to be on its side, so I strap my head to the side of a sleeping bag and say, 'Are you happy now?' "

Kathy Thornton said, "When it was time to go to bed, Dick covered the windows and sent the kids down to the middeck to go to sleep." The crew slept peacefully for several hours while computers and watchful eyes at Mission Control monitored events on board as *Endeavour* orbited Earth.

CHAPTER SIX

A WALK IN SPACE

Early on the third day, *Endeavour* was rapidly catching up with the telescope. The crew was excited. Just before sunset, they spotted the Hubble. It was a beautiful sight, glowing blue and silver in the fading light. "Now it's all eyeballs and hands," Sox said, as he and Commander Dick Covey took over from Mission Control and fired small burns from the maneuvering rockets. Slowly and gently, the shuttle approached the forty-three-foot-long telescope looming outside its windows. The rocket burns used fuel, and Ken Bowersox was worried. He knew that there was only one chance to capture the Hubble. If they weren't in the right position, there wouldn't be enough fuel for a second attempt. If they didn't grab the Hubble on their first and only try, the mission would have failed before it began.

Claude was ready at the controls of the robot arm, waiting for the right moment. *Endeavour* glided into place below the telescope. Claude reached out with the mechanical arm and snared it! The telescope was safe, snugly tucked into the payload bay. With relief, Dick Covey radioed Mission Control. "Houston, *Endeavour* has a firm handshake with Mr. Hubble's telescope. It's quite a sight." The first crucial step of the repair mission was over. It was time for the space walks to begin.

Now the first surprise took its "bite." One of the solar arrays was badly twisted. Would it roll up as planned? Story and Jeff would look more closely the next day, when the "house calls" on Hubble would begin. Commander Dick Covey summed up the spirit of the crew when he said, "We are ready. We

While the Hubble Space Telescope was being designed, Story Musgrave suggested that it be fitted with handrails to make it easier for spacewalking astronauts to service it. Story uses those handrails on his first space walk.

are inspired. Let's go fix this thing."

The next day, Kathy and Tom helped Story and Jeff put on their space suits. Their very lives would depend on them. Outer space is just that—space. It is a void. There is nothing in it—no oxygen, no hydrogen, no carbon dioxide—nothing. Human beings without protection would die in that environment in a matter of seconds. The air would rush out of our lungs. Our blood would boil, our skin expand. And as if that weren't a terrible enough picture, we would roast or freeze, depending on whether it was day or night.

Going out into space is not like going out your front door. Story and Jeff entered and left the shuttle through its air lock, a small chamber with two hatches, or doors. Before leaving the shuttle, they had to breathe pure oxygen for about forty minutes in order to get rid of the nitrogen in their bloodstream. In outer space, nitrogen can cause a painful and dangerous condition known as the bends. After their "pre-breathe" was over, all of the air was pumped out of the air lock. Now

On the fifth space walk, Story Musgrave and Jeff Hoffman performed final repairs. Here they prepare to put new covers on the telescope's magnetometers.

it was a vacuum, just like space. The astronauts could safely open the outer door and float into the payload bay. "It's an exciting moment when you open the air lock and see the entire universe staring you in the face," Jeff said, as he and Story started the first space walk.

For safety, Story and Jeff each attached themselves to the shuttle with a tether, a long metal cable. Moving deliberately and cautiously, Story pulled himself hand over hand along the sixty-foot length of the payload bay to the waiting telescope. His feet secured in foot restraints, Jeff got a ride on Claude's robot arm. All went well as they replaced the faulty gyroscopes—until they tried to close the doors to the instrument compartment. A second surprise took a "bite."

One door sagged slightly below the other, and they would not close. Jeff and Story struggled with the balky doors. Because they no longer matched, the closing bolt would not slide home. The astronauts used their power tools with no results. They talked to Mission Control and even considered leaving the doors slightly ajar. But Story was determined. In a move that was not rehearsed ahead of time, he went to the toolbox and removed a tool called a come-along. The come-along held the two doors together, leaving Story's hands free to push on them and force the closing bolt home. The operation was a success, and Story and Jeff earned a new nickname from NASA—Mr. Goodwrench.

Next, the crew turned their attention to the twisted solar array. The solar arrays were designed to roll up like window shades. The undamaged array rolled up smoothly for storage, but the other would not roll up because it was warped. The Even Couple—Kathy and Tom—would have to throw it overboard the next day, before installing the new arrays.

Story and Jeff put Kathy and Tom in their space suits for the second space walk and sent them off to work. Kathy rode the robot arm while Tom floated. The damaged solar array was still producing electricity from the sun, so Kathy and Tom had to work during the orbiting "night." They had lights on their space helmets, like miners, and from inside the shuttle, Story helped by shining a spotlight on the work area. Still, it was hard for them to see exactly what they were doing. Kathy steadied the large floppy array while Tom unfastened it.

The solar arrays were attached to the telescope with electrical connectors similar to the plugs on the back of a computer. Each connector had many small pins that could be easily bent—especially by someone trying to plug or unplug them wearing bulky space gloves! The connectors and bolts were essential to the telescope's life. If they were damaged during this repair, the new arrays could not be installed. Without solar energy, the Hubble would have no electricity. Without electricity, the Hubble would have no life.

The team worked slowly and deliberately as *Endeavour* fell around the Earth. At last, Kathy, riding on the end of the robot arm, was ready for the crucial and dangerous moment when she alone would hold the array. She had to keep it steady and still. If it flopped around it could hit and damage the telescope. The

Perched on the end of the robot arm, Kathy Thornton steadies the damaged solar array while she waits for dawn and the right moment to release it.

Hubble had enough trouble; Kathy did not want to "break what isn't broken," as Jeff would say. She waited for Tom's instructions.

"Coming out real smooth," Tom said. "Four inches . . . six inches . . . eight inches out." His voice was calm, a gentle coach.

Kathy said, "OK, Claude, real easy." She truly had her hands full, as the robotic arm pulled her slowly away from the telescope.

"OK, K. T. I'm letting go," said Tom.

"I have it," Kathy answered. The array was perfectly steady and Kathy exclaimed with delight, "Holy moley, a piece of cake!"

But tension mounted again. Everyone was worried about the moment when Kathy would toss the array overboard. It had to be done in daylight, when the entire crew could see clearly. No one wanted the array to hit the top of the Hubble on its way to becoming space junk. Poised on the end of the robot arm, Kathy held it over her head and waited for the sun to rise over Africa.

"I think I see sunrise coming, K. T.," Tom said.

At last, the word came from Story, inside the shuttle. "OK, Tom. Tell K. T. to go for release."

Tom had one request. "Can you just hang on one second, so I can get to where I can watch?" He moved to get a better view, then said, "OK, K. T. You ready?"

"Ready."

"Got a go for release," Tom said. The moment had come. "OK. No hands!" Kathy said as she let go of the 350 pounds of useless metal and fabric.

"There it goes," said Tom. The panel revolved and drifted away. Sox fired a few small burns from the shuttle's rockets to move *Endeavour* and the telescope out of danger. In the void of space, the maneuver caused an artificial wind.

Kathy's voice came over the radio again. "Wonder what it's going to do when it starts flapping in the breeze?"

Suddenly, as she saw the great orange array flexing over the desert, she exclaimed, "It's almost like a bird, Tom. Look at it!" They stared at the incredible sight of a giant "bird" soaring over the African desert below.

Jeff Hoffman rides the robot arm and prepares WF/PC II for its installation in the telescope.

The results of all the planning and practice were becoming more and more obvious as each team of astronauts completed their space walks successfully. On the third space walk, Story and Jeff installed the new camera, WF/PC II. The critical moment came when Jeff held the camera steady and Story removed the protective cover on the mirror. One slight touch or bump, and the mirror could have been contaminated or knocked out of alignment, a catastrophe that would have crippled the telescope. The astronauts' movements were slow, precise, and delicate. Everyone relaxed a bit when that job was finished, and Jeff said excitedly, "I hope we have a lot of eager astronomers ready to use

Because COSTAR had to be held directly in front of her face, Kathy Thornton relies on Tom Akers's instructions while guiding it into place.

this beautiful thing."

On the fourth walk, Kathy and Tom became orbiting eye doctors and slid the refrigerator-sized COSTAR into the telescope. Kathy moved the six-hundred-pound instrument with ease, but COSTAR was in front of her face. She knew she had only millimeters of clearance, but she couldn't see what she was doing. Once again, her partner became her eyes, and COSTAR slid into place without a hitch. Tom hummed while he worked. Later, Claude recalled that sound and said, "It was good to hear Tom humming, because we knew when Tom was humming things were going well. And Tom was humming most of the time!"

When the Hubble's "glasses" were

in place, Kathy said, "I think everyone can breathe a sigh of relief." Jeff added, "It should be exciting to see what the Hubble can do with a new set of eyeballs."

On the fifth EVA, when Story and Jeff were winding up the repairs, a tiny screw came loose and floated away. On Earth, a dropped screw is not a disaster. In space, it could be. It could contaminate the Hubble. Jeff and Claude spotted it at the same time; Claude noticed it even from his position inside the shuttle. The "famous screw chase," as Story called it, was on. Jeff rode the arm while Claude steered him toward the floating screw, in hot pursuit. Later Jeff said, "When we were chasing it, I actually felt like a kid riding on a merry-go-round, going after the brass ring, holding on with one hand and reaching out." Jeff got the "brass ring," and he and Story finished up their work. They did a final cleanup of the payload bay and reentered the shuttle, knowing that the Hubble was now ready to go. The time had come to send it on its way.

Claude raised the towering observatory from its workbench and released it into orbit once again. As it moved into the blackness of space, its orange solar arrays glowed in the sunset. "It will look far into the cosmos and far into the past," he said. "It is a time machine as well as a space exploration machine."

Looking out the shuttle's window, Tom was wistful. "The payload bay really looks empty."

In a record-breaking five space walks—no mission to space had ever had that many—the heroes of STS-61 had accomplished what many said could not be done. It was time for these space mechanics to return to Earth.

(Right) Repaired and ready to observe the universe, the Hubble Space Telescope floats away from Endeavour*'s cargo bay.*

Kathy Thornton works on COSTAR. A checklist containing vital information was attached to the sleeve of each astronaut's space suit.

CHAPTER SEVEN

HOMECOMING

"At the end of each day we had celebrations of joy and relief that we had reached each milestone, but we saved the big celebration for the very end," Commander Dick Covey said.

There was some good-natured kidding as the astronauts shared their thoughts about viewing Earth from outer space. Tom Akers recalled looking down on the United States at night and seeing the country stretched out below him from the Pacific Ocean to the Atlantic. The cities were brightly lit, and he tried to pick out Eminence, Missouri, but he could not spot it. He commented, "Well, it's eleven-fifty at night. There's probably nobody up in Eminence." One of the crew members answered, "Tom, it wouldn't matter if all five hundred people in Eminence turned on their lights—you *still* wouldn't see it!" Though he was unable to see his small hometown from outer space, Tom Akers said, "I feel real lucky that I got to see the earth from up here."

Sox took a last look at the sight of Earth from the shuttle. "You don't see any borders," he said. "It makes you feel like you're a citizen of the planet, not just a citizen of your country."

The crew's trainers had provided "surprises" inside their personal lockers, and now was the time to look at them. One was a Dr. Seuss book that had a new title and an altered text that made jokes about the astronauts. Instead of *One Fish, Two Fish, Red Fish, Blue Fish,* the

The happy crew of STS-61 takes a moment to celebrate the success of its mission before preparing for the journey home.

trainers renamed the book *One JIS, two JIS, red JIS, blue JIS* to remind the astronauts of all six of the full-dress rehearsals—joint integrated simulations—that everyone had endured as they prepared for this mission.

To celebrate the season, Jeff Hoffman had brought along a traveling menorah and a dreidel to observe Hanukkah, the Jewish Festival of Lights.

The menorah is a candelabra with places for nine candles, and the dreidel is a small top that is traditional to the holiday. Jeff spun the top and laughed. "I'm trying to decide how we might reinterpret the rules for spaceflight, since there's no up or down here." Then he added, "We won't light the candles in the spacecraft, but with my silver traveling menorah and the dreidel, I'm all set for my onboard celebration."

Kathy Thornton said, "It seems very appropriate that this is the season of miracles. We had our miracle when we got through this mission as easily as we got through it."

There were phone calls from the president and vice president of the

United States. President Bill Clinton said this flight was "one of the most spectacular space missions in history." Vice President Al Gore added, "The whole world will see the space program once again as a symbol of the highest aspiration of humankind."

The joyous mood continued as everyone basked in the glow of knowing that they had worked their hardest and done their best. The surgery had been a success. But had the Hubble been cured? The world would have to wait weeks for the answer, while scientists fine-tuned the telescope.

After eleven days in space and a journey of 4.4 million miles, it was time to return to Earth. It was time for Tom to give his wife, Kaye, her golden charm in the shape of the mission's insignia, and David and Jessi the watches that had flown in space. It was time to come home to family and friends. It was time, as Kathy said, "to fly a desk for a while."

Once again, rockets ignited, but this time they slowed the shuttle down, so it could leave Earth's orbit and reenter the atmosphere. As *Endeavour* approached Earth, Dick Covey once again took over the controls. There was no motor; the space plane had become a glider. Silently and in the dark of night, it landed and returned to its home at the Kennedy Space Center. "Moments of great adventure come once in a lifetime," Commander Covey said. This had been one of them.

Touchdown at Kennedy Space Center. Mission accomplished.

AFTERWORD

Scientists measure the incredible distances in our universe in terms of light-years, rather than miles. A light-year is the distance that one ray of light travels in one year in a vacuum. Since light travels at 186,000 miles per second, one light-year is 5.9 trillion miles. Our solar system is approximately 25,000 light-years from the center of our Milky Way galaxy.

The Hubble Space Telescope has always been able to see objects that are 4 billion light-years away. However, on January 14, 1994, NASA scientists announced that the Hubble Space Telescope repairs were a success. Now it was able to see objects 12 billion light-years away—almost to the edge of our universe and the beginning of time. In other words, the Hubble can capture the image of an object as it appeared 12 billion years ago. COSTAR and WF/PC II worked!

Dr. Story Musgrave has said, "It is the nature of humans to be exploratory and to push on. It is the essence of human beings to try to understand their universe and to try to participate in the entire universe and not just their little neighborhood."

When the Hubble's troubles developed, scientists did not turn back from their goal of building a great Earth-orbiting telescope that would help them better understand our universe. Like all great explorers, they pushed on and did not let obstacles get in their way. Careful planning, determination, cooperation, and plain hard work among scientists, astronauts, and ground crews solved the Hubble's troubles. "Before and after" photographs demonstrated the truth of NASA's claim that the Hubble Space Telescope was "fixed beyond our wildest expectations." However, perhaps the spectacular results were best described by one astronomer from ESA who looked at the new Hubble photographs and exclaimed, "All I can say is, 'Wow!' "

(Top left) Before: A star photographed by the Hubble Space Telescope before the repair mission. The halo surrounding the star is out-of-focus starlight, caused by the spherical aberration problem in the telescope's primary mirror.

(Bottom left) After: COSTAR's tiny mirrors allow the starlight to be focused properly and give astronomers much clearer photographs to study.

Opposite page:

(Top right and far right) Before and after COSTAR: These are photographs of the central region of galaxy NGC1068, located 60 million light-years from Earth. By studying the "after" image, scientists hope to gain new understanding of the black hole they assume is at the galaxy's center, generating the knots and streamers of near-nuclear energy shown in this photograph.

(Bottom right and far right) WF/PC I vs. WF/PC II. Before and after pictures of the nucleus of distant galaxy M100: The first image was taken by WF/PC I on November 27, 1993. The second image was taken by WF/PC II on December 31, 1993. Like our Milky Way Galaxy, M100 is spiral shaped. It is located tens of millions of light-years from Earth, but the Hubble Space Telescope now gazes clearly upon it.

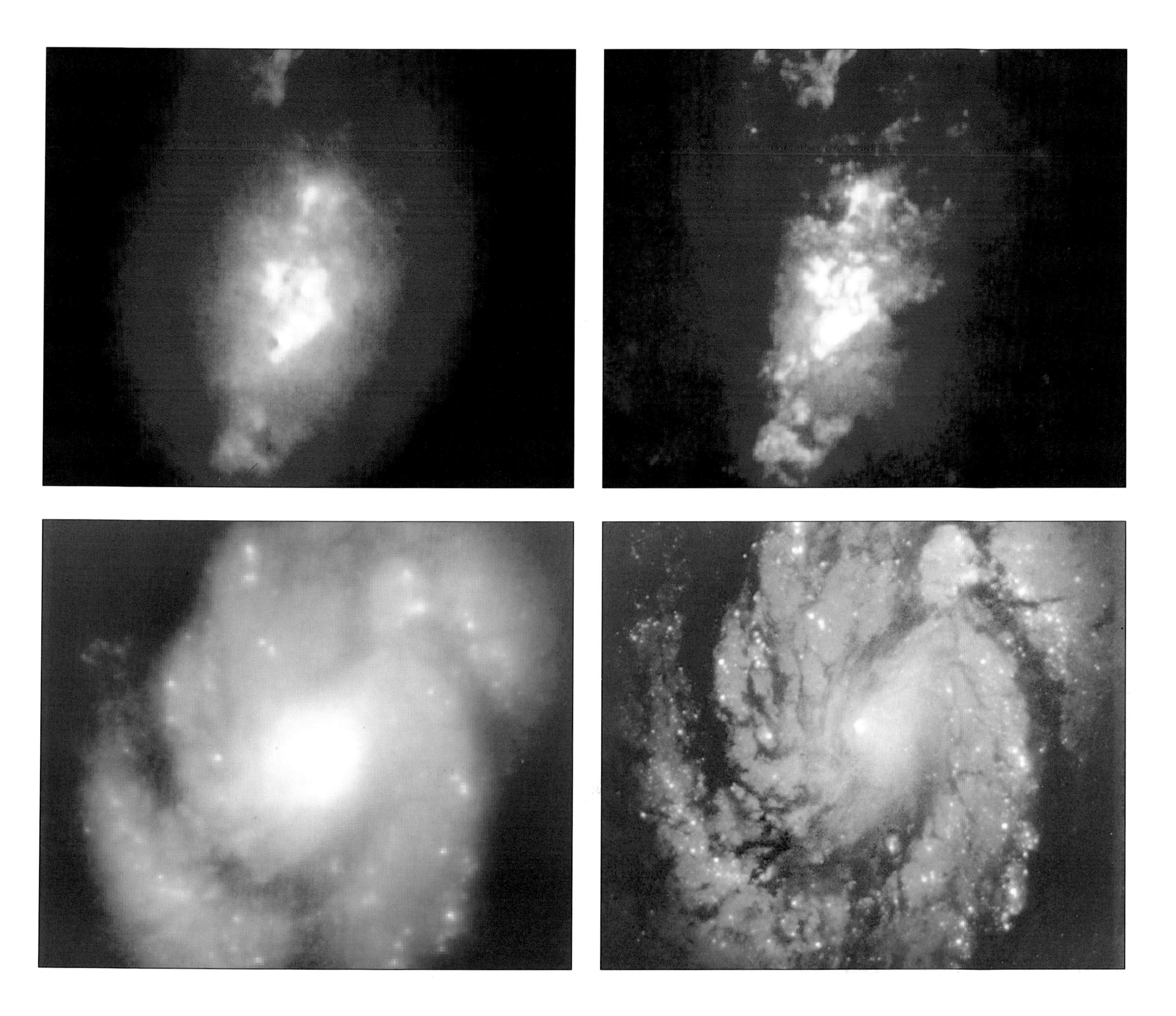

NEW HANOVER COUNTY PUBLIC LIBRARY
3 4200 00397 2680